Prescription to My Younger Self:
What I Learned After Pharmacy School

Alisha Broberg

Jennell Colwell

Brad Koselke

Annah Steckel

Erin Albert

AuthorHouse™
1663 Liberty Drive, Suite 200
Bloomington, IN 47403
www.authorhouse.com
Phone: 1-800-839-8640

© 2008 Pharm, LLC. All rights reserved.

No part of this publication may be reproduced, stored in a retrieval system, or transmitted in any form or by any means, electronic, mechanical, photocopying, recording, scanning, or otherwise, except as permitted under Section 107 or 108 of the 1976 United States Copyright Act, without the prior written permission of the CEO of Pharm, LLC, Dr. Erin Albert.

Limit of Liability/Disclaimer of Warranty: While the publisher and author have used their best efforts in preparing this book, they make no representations or warranties with respect to the accuracy or completeness of the contents of this book and specifically disclaim any implied warranties of merchantability or fitness for a particular purpose. No warranty may be created or extended by sales representatives or written sales materials. The advice and strategies contained herein may not be suitable for your situation. You should consult with a professional where appropriate. Neither the publisher nor authors shall be liable for any loss of profit or any other commercial damages, including but not limited to special, incidental, consequential, or other damages. The information contained herein is not necessarily the opinion of the authors or the publisher.

Designations used by companies to distinguish their products are often claimed by trademarks. In all instances where the author or publisher is aware of a claim, the trademarks have been noted where applicable. The inclusion of a trademark does not imply an endorsement or judgment of a product or service of another company, nor does it imply an endorsement or judgment by another company of this book or the opinions contained herein.

First published by AuthorHouse 3/20/2008

ISBN: 978-1-4343-6259-9 (sc)

Printed in the United States of America
Bloomington, Indiana

This book is printed on acid-free paper.

In memory of J. H.,

*and to the faculty, staff, and students of
Butler University's College of Pharmacy and Health Sciences*

Table of Contents

FOREWORD xi

OATH OF A PHARMACIST xiii

LETTERS FROM CONTRIBUTING PHARMACISTS

Erin Albert 3
   MBA, PharmD, Assistant Professor and Director of the Ribordy Center for Community Practice, Butler University COPHS

Mary Andritz 7
   PharmD, Dean Butler University COPHS

Anonymous 11

Scott Baker 15
   RPh, Senior Vice President of Stores East, CVS

Bernadette (Bonnie) Brown 19
   PharmD, Assistant Dean for Student Affairs, Butler University COPHS

Jack Devine 23
   RPh, Retired

Denise Leonard Dickson 29
   PharmD, MBA, Director Six Sigma, Eli Lilly & Company

Joseph DiPiro 33
   PharmD, Executive Dean, South Carolina College of Pharmacy

Ken Fagerman 39
   RPh, MM, Clinical Pharmacist/Manager, Infusion Pharmacy, St. Joseph Regional Medical Center

Nathan Gabhart 45
    RPh, Vice President of Pharmacy Operations, Independent Pharmacy Chain

Jane Gervasio 49
    PharmD, BCNSP, Director of Clinical Research and Scholarship, Butler University COPHS

Judith Jacobi 53
    PharmD, FCCM, FCCP, BCPS, Critical Care Pharmacy Specialist, Methodist Hospital

Julie Koehler 59
    PharmD, Associate Professor and Chair of Pharmacy Practice, Butler University COPHS, Clinical Pharmacist, Clarian Health Partners

Lindsay Koselke 65
    PharmD, Clinical Pharmacist, St. Anthony Memorial Hospital

Lucinda Maine 69
    PhD, RPh, Executive Vice President/CEO, American Association of Colleges of Pharmacy

Susan Malecha 73
    PharmD, MBA, Senior Medical Science Liaison National Manager, Genentech

Rhalene Patajo 79
    PharmD, Manager of Regional Medicine, Boehringer Ingelheim Pharmaceuticals, Inc

Amanda Place 83
    PharmD, Pharmacy Manager/Co-Owner, Access to Care Pharmacy

Jeffrey Rein 87
    RPh, Chairman and CEO, Walgreens

Denis Ribordy 91
    RPh, Founder, Ribordy Drugs

Byron Scott 95
    RPh, Senior Director Regulatory Affairs, Schwarz Biosciences, Inc

Ron Snow 101
    RPh, Manager, Professional and College Relations, CVS

Bill Sonner 105
    RPh, Divisional Director of Pharmacy Operations, Walgreens

Marilyn Speedie 111
    PhD, Dean, College of Pharmacy Professor, Department of Medicinal Chemistry, University of Minnesota

Leticia Van de Putte 117
    RPh, Texas State Senator

John Watt 123
    RPh, IV Therapy Program, Acute Care for the Elderly Team, Consultant Pharmacist, SNU Unit, Wishard's ECF

Hanley Wheeler 127
    RPh, Senior Vice President of Central Operations, CVS/Caremark

LETTERS FROM CONTRIBUTING STUDENTS

Alisha Broberg 133
    PharmD Candidate, Butler University

Jennell Colwell 137
    PharmD Candidate, Butler University

Brad Koselke 141
    PharmD Candidate, Butler University

Annah Steckel 145
    PharmD Candidate, Butler University

ACKNOWLEDGMENTS 149

Foreword

As current P4 (final year) students at Butler University, the four of us began a research project in hopes to find answers for the questions - what happens in terms of our education after pharmacy school, and what can we learn from our mentors already in practice? Rather than performing a retrospective chart review or a prospective survey, we instead choose to co-author/co-edit a book, *Prescription to My Younger Self: What I Learned after Pharmacy School*.

This book is a collection of letters written by pharmacists in varying professional settings, who we feel have either excelled in the profession of pharmacy, or who have inspired us and were willing to share their stories in print. We asked the pharmacists to provide a biography along with a letter they have written in the present to their former self on the day that they graduated from pharmacy school, and share the pearls of wisdom they learned about their profession after graduation. Included in the epilog is group of letters from a different perspective- the student. Each of us contributed a letter reflecting upon our current aspirations and hesitations as we venture into the unknown, a career. *Prescription to My Younger Self: What I Learned after Pharmacy School*, allowed us to step outside of the box and take a chance on a very unique and rewarding project.

Throughout this journey, we not only conquered the process and hardships of publishing a book, but we discovered many unknown strengths that each of us encompasses. It was a battle to overcome many of our own personal hesitations such as: writing fears, contacting professionals, and hearing the answer "No". This book contains many wonderful letters written by pharmacists we contacted proving that it never hurts to ask.

Finally, the contributing pharmacists have bestowed upon us exceptional advice on how to be successful and be content in the future. Each pharmacist has contributed greatly to the profession and we are extremely grateful for his or her participation and shared wisdom. The letters featured in our book reiterate that possibilities are infinite following graduation.

Oath of a Pharmacist

At this time, I vow to devote my professional life to the service of all humankind through the profession of pharmacy.

I will consider the welfare of humanity and relief of human suffering my primary concerns.

I will apply my knowledge, experience, and skills to the best of my ability to assure optimal drug therapy outcomes for the patients I serve.

I will keep abreast of developments and maintain professional competency in my profession of pharmacy.

I will maintain the highest principles of moral, ethical, and legal conduct.

I will embrace and advocate change in the profession of pharmacy that improves patient care.

I take these vows voluntarily with the full realization of the responsibility with which I am entrusted by the public.

Developed by the American Pharmaceutical Association Academy of Students of Pharmacy/American Association of Colleges of Pharmacy Council of Deans (APhA-ASP/AACP-COD) Task Force on Professionalism; June 26, 1994.

Letters from Contributing Pharmacists

Erin Albert

Erin Albert *is the Director of the Ribordy Center for Community Practice at Butler University in the College of Pharmacy and Health Sciences, and teaches as an Assistant Professor. She also works part time as a PRN pharmacist in community practice, owns two companies, and has written two books. Dr. Albert has worked in several industry and other practice settings including: retail pharmacy, clinical research operations, consulting, drug safety & epidemiology, medical affairs, and pharmaceutical brand/marketing. She obtained her BS in pharmacy from Butler University, her MBA in Marketing from Concordia University Wisconsin, and her Doctor of Pharmacy (PharmD) from Shenandoah University. She graduated from pharmacy school with her BS in the early 1990s.*

Dear Erin,

Congratulations! I know what you are thinking: getting through pharmacy school was one of the toughest challenges you've had thus far. Trust me when I say, there will be even more and bigger challenges ahead. Here is what you should keep in mind as those challenges arise:

Things are not always black and white, which is very UNLIKE pharmacy school. This will be a tough lesson for you to learn. In school you were taught one best way to answer or solve problems. However, this is not frequently the case in the school of hard knocks (aka life). Remember this, as it will help you deal with all those future decisions you will make that will not be 100% right or wrong. Welcome to the art of living.

Also, know that you can create your very own unique and special career path tailored like couture, made only for you. You will get frustrated with your work. You will get bored as well. You will have managers and bosses with whom you will not always agree. However, as you move through your career, you will forge your own path, and in the forging alone you will receive great professional and personal rewards. Don't be afraid to create something if it doesn't exist. Just like one of your favorite quotes - Gandhi said, "Be the change you wish to see in the world." You will not always have the answers to everything right away, but try and trust your gut and go for it…be bold and live passionately. Failure is failure only if you learned nothing from the experience.

Keep learning. Just because you've graduated from school does not mean your brain will go on permanent holiday. You enjoy learning, so learn from everyone and everywhere if and when you can; education inside and outside the classroom will make your career and life even more fun!

Speaking of fun, you will find that the more you help others along their journey, the more rich, rewarding, and fun your own professional journey will be. Focus on other people's needs first. Give back time, money, and your talent. Many people along the way have helped you along your path, and you owe it to the universe to try and make it a better place by passing your blessings on to others.

Finally, while you are busy making the universe a better place, don't forget to have some fun beyond work every once in awhile. Getting your first job in the real world and managing your adult life is very serious work, granted. But seriously having fun is also important and key to a great life and career. If your work can't be described as fun anymore…move on to another challenge.

Cheers,

e

Mary Andritz

Mary H. Andritz *is the Dean and Professor of Pharmacy Practice at Butler University College of Pharmacy and Health Sciences. She received a Doctor of Pharmacy degree from the University of New York at Buffalo and a BS in Pharmacy from the University of Pittsburgh.*

Previously Dr. Andritz held positions as the Vice-President for Clinical and Professional Development, Dean and Assistant Dean for Professional Affairs at Albany College of Pharmacy (ACP). She was a tenured faculty member in Pharmacy Practice at ACP with a clinical practice in infectious disease. Other positions held by Dr. Andritz include Executive Director of the New York State Council of Health-system Pharmacists (NY state affiliate of American Society of Health-System Pharmacists-ASHP) and Chief of Pharmaceutical Care and Residency Director at the Stratton Department of Veterans Affairs Medical Center in Albany.

Dr. Andritz received the Practitioner Recognition Award (FASHP) for sustained contributions to the profession from the American Society of Health-System Pharmacists in 1998 and the Distinguished Alumni Award from the University of Pittsburgh in 2005. Lambda Kappa Sigma International Pharmacy Fraternity honored her for distinguished achievement in the Pharmacy profession with their Award of Merit in August 2004.

Dear Mary,

You've worked hard and deserve to be proud of becoming a pharmacist! You've come to understand that your actions reflect not only on you but also on your profession. Sometimes you will be asked to be a spokesperson for pharmacy, but every day you can be the "face" of the profession to someone with whom you interact.

The success that you had as a student was linked to long hours of studying and the pursuit of excellence. You will have to add other strategies for success when you no longer have the optimal amount of time to devote to a task. "Compromise" may not come naturally to you, but it can become an indispensable tool in selected situations.

Pay attention to your strengths and to those of others. You'll notice that it's easier for you to do some things; your colleagues might have different strengths. While you may be tempted to spend considerable time improving your "weaknesses," it is often feasible to partner with others who have complementary strengths. That way each person gets to do what they do best! Your contributions to successful teams in the future will be influenced by the knowledge, skills and attitudes that you will develop through life experience. Don't presume to know where your career will take you. Be open to the possibilities!

While you may expect to continue learning as part of your career path, you may be surprised how much development occurs through non-work challenges. When you "stretch" beyond your comfort zone, you can grow in ways that you might not expect.

You've reached graduation, an important milestone. Don't forget to enjoy the journey to other milestones in your life. Some of the most precious moments in life occur while you're waiting for the "big events."

Best wishes for the wonderful journey in front of you!

Mary

Anonymous

The following pharmacy professional chose to remain anonymous.

Dear Anonymous:

Here are a few things to consider now that you are embarking upon your career in pharmacy:

1. Learn right now about finances, how money compounds and why you should begin to invest now for your future.

2. Do some research and decide where you want to live. Move there now, get established and move up. Perhaps you don't want to live in one place for the rest of your life – decide now and conceive a plan.

3. There are lots of workplaces in pharmacy and lots of open jobs. Take a job that will make you happy, that you will leave at the end of the day feeling good about your work.

4. Don't be hasty about getting married. Get to know this person very well before you tie the knot. Are you really a fit? The personality and character may say a lot about what your spouse will be like or what baggage they will carry. You might invest a few dollars and have a good psychologist assess your compatibility.

    Will this person allow you to be who you are and want to be? Are you able to do the same? Do they want to live in the same size and type of community as you and in the same area of the country? Discuss these things and children, money, jobs and other topics openly and at length well ahead of time. There are lots of people to love but few that truly fit with you.

5. Consider graduate school very carefully before making that investment of a few up to twelve years. It means lost wages, lost retirement funds, time away from your family while

you study and very possibly no additional earning power as compared to the PharmD degree that you are about to earn.

6. In the workplace, that advancement you have been offered may sound impressive. But give careful thought to your change in relationships with those you work for. Would you rather be their friend than their boss? Think about the change in the work you will do. Would you rather be on the floor doing clinical pharmacy or dealing with financial reports, trying to get needed resources from hospital administration and dealing with employees who are reluctant to change? Being a supervisor/boss can be a killer, emotionally, physically and in terms of burnout.

7. Enjoy life, travel or whatever you enjoy doing. Seek lots of quality of life early. Spend lots of time with your kids. It is good for them in several ways and for you. Truly, before you know it they will be grown and those memories will make you feel fulfilled and content in your rocking chair.

Sincerely,

Anonymous

Scott Baker

Scott Baker *is the Senior Vice President of Stores East for CVS Pharmacy. He is responsible for the operations of nearly 2000 drug stores along the East Coast. His career started with Peoples Drug as a pharmacist/store manager in the Indianapolis area and has worked several field management positions in the retail drug business with Reliable, Revco and now CVS Pharmacy. Scott obtained his BS in pharmacy from Butler University in 1982.*

Dear Scott,

Great Job! You should feel good about graduating from pharmacy school. You have put a lot of hard work into your achievements, but I assure you it will take an even higher level of effort to transition from school into the mainstream of life. To help you move ahead, I thought I would share with you some "Graduation Advice I Never Got…but wish I had":

Spend time with people who aren't like YOU! Real diversity is difficult to find and I don't just mean the gender/racial kind. Only now, after college, can you really benefit from diversity. Start by getting to know people whose skin is a different color from yours. Surround yourself with friends of every political opinion and listen to them RESPECTFULLY. Keep broadening your horizon by seeking out real diversity.

READ for Pleasure. This may strike you as odd advice since you've been reading books for years. Now that you're free of that college studying….the last thing you probably feel like doing is sitting down with a good novel, much less a serious magazine or newspaper and that's a shame. As a college graduate you are on top of the WORLD. But there's only one way to stay there. READ! That simple fact separates the leaders from the followers, achievers from the rest of the crowd. It's a truism, because it's true, knowledge is power and the habit of reading is the path of knowledge.

Choose Reality over Theory. There is nothing wrong with theories. Theories can be interesting, as you know, having spent your life so far immersed in them. Now that you're leaving college, you're going to have to become a realist. What matters now is what WORKS! Getting to what works, sometimes means accepting a little failure. These lessons will serve you well as you innovate the next solution.

You "create" the environment for SUCCESS, take advantage of this fact. Whether it is how you practice pharmacy or live your life, you control your attitude on how you approach the day. Many words have been written about the power of attitude and it is easy to understand why. A positive attitude causes a chain reaction of positive thoughts, events and outcomes, creating extraordinary results.

A final thought, one that you're heard countless times. Unlike some platitudes, this one is on the money. Your life will be half over before you know it. It may already be. None of us can know. So don't squander a moment. Get up, get moving and good luck!

Regards,

Scott

Bernadette (Bonnie) Brown

Bernadette (Bonnie) Brown *(BS Pharmacy 1979, PharmD 1981 Purdue University) is currently the Assistant Dean for Student Affairs and an Associate Professor of Pharmacy Practice at Butler University College of Pharmacy and Health Sciences. In this position, Bonnie is responsible for the College's student organizations, student advising, and coordination of major events. Previously, she worked as a Clinical Pharmacist for 18 years in a variety of settings around the Indianapolis area, primarily health-systems and as a consultant pharmacist. Her primary area of interest is Geriatrics.*

Dear Bonnie;

Finishing pharmacy school is a wonderful accomplishment, but just the beginning. Life definitely does not get easier. Trying to balance family and career is the hardest juggling act you will ever do. Hopefully these tips will help you in life:

- Family comes first. Cherish your husband and children. Nothing is more important than your life with them. Keep in close contact with your brothers and sister, you shared so much with them growing up; continue to grow old with them.

- Do not neglect your spiritual life. It can get you through the tough times, which are sure to come.

- Find the right job. You always wanted to be a teacher, but instead chose the pharmacy path. Keep teaching while being a pharmacist and you may land up in academia after all. It is a very rewarding career choice, and allows the balance of family and career better than many jobs.

- Give back to the profession. Stay involved with local, state, and national organizations. Seek out ways to serve the profession and the public.

- Continue to learn. Not just for continuing education, but to broaden your horizons. Keep up with the world and be an advocate for issues you believe strongly in. Read a book just for fun once in a while!

- Keep healthy. High blood pressure and heart disease run in the family. Take care of yourself, exercise, eat well, and learn to relax. This will allow you to live a long life and enjoy your later years.

- Be a positive role model for future generations of pharmacists. Confucius said, "When walking in a party of three I always have teachers. I can select the good qualities of the one for imitation and the bad ones of the other and correct them in myself."

Always enjoy the best life has to offer,

Bonnie

Jack Devine

Jack Devine *is the founder and former owner of a very successful community pharmacy in Metamora, IL. Presently, he works part time at his old store. He graduated from St. Louis College of Pharmacy in 1956, where he was a member of the Rho Chi society. His family includes his wife Sue (RN), who he has been happily married to for 51 years, and 3 children (1 is a pharmacist). Special honors include: Bowl of Hygeia award in 1996, Central Illinois Pharmacists Assoc. Hall of Fame award in 2002, and StLCoP mortar & pestle society lifetime member in 1998.*

Dear J,

Congratulations! You certainly picked one of the most respected professions, because pharmacists help people and society every day. Two reasons come to mind: (1) we are very accessible and (2) we charge little or nothing for our advice.

Be prepared to continue learning. This is a lifelong commitment. We are and need to continue to be the drug experts. Physicians and our patients turn to us for advice and knowledge. At the time I graduated, sulfonamides, aureomycin, penicillin G and chloromycetin were the anti-infective choices. Meprobamate had just come out and was the hot tranquilizer drug. Dilantin® and phenobarbital headed up the anti-seizure medications. As you know, drug therapy is now much more complex and advanced and will continue to be. Fifty years ago if a patient had a question about their medication you were told to answer, "You need to ask your doctor about that." There was mystery in medication and you were not allowed to answer. Times have changed for the better.

Be sure to join your local, state, and national pharmacy associations such as APHA, NCPA, and ASCP. They work for the betterment of you and the patient. If possible, attend their annual meetings. It's a great place to obtain CE and network with your colleagues.

The patient always comes first in our world. Without them we are not needed. I always set goals, and when I graduated from pharmacy school I wanted to own my own pharmacy. I always tried to serve my patients and be available to answer their questions. If you do that, they will be intensely loyal to you.

As you will learn as your career advances, our profession is greatly impacted by the government. The National Community Pharmacists Association has a saying: "Get into politics or get out of pharmacy." Support your state and federal elected officials and make your views known to them on pharmacy issues. They will appreciate that. By support, I mean not only by your vote but also monetarily through your pharmacy association's political action committee. As you mature you will learn that forces are out there that are not always friendly to pharmacy.

Your family comes first, and your children will grow up and leave home before you know it. It is important to spend quality time with each of them every day. Enjoy and learn from them. I know it is hard to do when you are required to work some evenings, weekends, and holidays, but it is important to try.

Give back to your community when you can. In my case, I was involved with the Boy Scouts and the American Red Cross in earlier years. I later served on our high school board of education for three terms and am still active in my church. I hope my small efforts made our community a better place to live.

Finally, don't forget your college of pharmacy. I know you paid your tuition, but it did prepare you well! Down the road, give back to them financially if asked. Also, if ever approached to be a preceptor (I was, but not from my college), please accept. I had pharmacy students for several years in my pharmacy. While helping them learn about community pharmacy, I felt that I learned more from them. I am proud that seven high school students that worked in my pharmacy went on to become pharmacists. Two of them now teach in colleges of pharmacy. One is my daughter.

You have a great future ahead of you with many career opportunities. Please make the most of it.

Best regards,

J

Denise Leonard Dickson

Denise Leonard Dickson *was born in Albuquerque, New Mexico, but has lived in nine states and Europe because of her father's career. She attended two years of college before being accepted into the PharmD program at University of California at San Francisco. She graduated from that program in 1979 and worked one year as a clinical pharmacist at University Hospital in San Diego. She went on to receive her MBA from the University of Pennsylvania, Wharton School of Business. Upon graduation in 1982, she joined Eli Lilly and Company in Indianapolis, Indiana, where she has spent her entire career. Over that time she has advanced and has held a wide variety of positions. Her assignments have provided her a deep understanding of the drug development process from discovery to commercialization. She has worked in multiple functional areas and with affiliates around the world. These experiences have taught her the challenges of drug development and the importance of meeting the needs of patients, physicians, payors and pharmacists. Denise married her husband, Dennis, in 1983. He started a company, Indy Tire Centers, in 1985, that has been very successful. They love the desert and have a second home in Tucson, Arizona. Their hobbies include travel, tennis and auto racing.*

June, 1979

Dear Denise,

Congratulations! You have a lot to be proud of as you complete your PharmD program. It has been a journey of ups and downs, but now it is behind you! Life is full of journeys and the next one is waiting just around the corner.

August, 2007

There are countless lessons I could share with you based on the paths that I have chosen in my life and in my career. However, as you know, nothing can compare to the lessons learned through your own life experiences. The decisions you make will dictate your future journeys. So, rather than a litany of my lessons I would prefer to share a brief verse with you which is called the Serenity Prayer.

> God, Grant me the serenity
> to accept the things I cannot change,
> courage to change the things I can,
> and wisdom to know the difference.

Remembering these simple words have provided me great inspiration over the years; I hope they will do the same for you as you choose your path and learn life's lessons.

My best wishes for a full and happy life.

Denise Leonard Dickson

Joseph DiPiro

Joseph T. DiPiro is Executive Dean of the South Carolina College of Pharmacy, which is an integrated program of the Colleges of Pharmacy at the University of South Carolina and the Medical University of South Carolina. Before coming to South Carolina, Dr. DiPiro was the Panoz Professor of Pharmacy at the University of Georgia College of Pharmacy and Clinical Professor of Surgery at the Medical College of Georgia. While there, he was also Assistant Dean for the College of Pharmacy and the School of Medicine at the Medical College of Georgia, and Head, Department of Clinical and Administrative Pharmacy. He received his BS in pharmacy from the University of Connecticut and Doctor of Pharmacy from the University of Kentucky. He served a residency at the University of Kentucky Medical Center and a fellowship in Clinical Immunology at Johns Hopkins University.

He is the Editor of The American Journal of Pharmaceutical Education. Dr. DiPiro has published over 120 refereed papers in academic and professional journals. His papers have appeared in Antimicrobial Agents and Chemotherapy, Pharmacotherapy, Critical Care Medicine, JAMA, Annals of Surgery, Archives of Surgery, American Journal of Surgery, Journal of Pharmacology and Experimental Therapeutics, and Surgical Infections. He is the senior editor for Pharmacotherapy: A Pathophysiologic Approach, now in its 6th edition. He is also the author of Concepts in Clinical Pharmacokinetics and Editor of the Encyclopedia of Clinical Pharmacy. He has served on the editorial boards of Pharmacotherapy and American Surgeon.

In 2002, the American Association of Colleges of Pharmacy selected Dr. DiPiro for the Robert K. Chalmers Distinguished Educator Award. He has also received the Russell R. Miller Literature Award and the Education Award from the American College of Clinical Pharmacy and the Award for Sustained Contributions to the Literature from the American Society of Health-System Pharmacists.

Dear Joseph,

It seems like ages since my graduation 30 years ago. I now have many years of hard-earned experience and insight that I hope you can use to make your career path clearer and easier than what most of us have traveled. Pharmacy has changed amazingly in the past 30 years. Some of the things we just talked about at graduation are now routine in pharmacy. Patient-oriented or clinical pharmacy has become the standard of practice. Pharmacists have proven that clinical functions benefit patients. You will find many more and different professional opportunities than I did after graduation. You have a rewarding career to look forward to. However, it may be quite different from what you expect right now. The profession will continue to change. As long as there continue to be great societal concerns with medications you will be needed. Being a problem solver will never leave you without a rewarding career.

In looking back over the years there have been a few surprises, mostly good ones. I am still working with people who I met 25-30 years ago, people who have become good friends and colleagues. Pharmacy is a small world. Some of the people who you meet in your first year after graduation will become lifelong associates. So, when you meet new people in the profession, consider that it may be the beginning of a 30-year or longer relationship. These individuals become your personal network and are very important to you to learn, grow, build new programs, seek advice, and extend your influence.

I have been fortunate that my career path put me in positions that allowed and even fostered personal and professional growth. I don't believe that all pharmacist positions do that. You should recognize that it is quite natural that your career and professional interests will change over time. Some organizations, these could be universities, health-systems, or corporations, or even your own business provide good opportunities for career growth.

Choose carefully for your first job, as a nurturing environment with mentors and career options is worth much.

Finally, don't be intimidated by big challenges. The big advances in our profession have come from people committing their effort over a long time. A modest effort over a long time can achieve great things. Recognize the excuses we all use to avoid big challenges: "I don't know how" "It would take too long" "It would be too expensive" or "What would people think of me if I did that." A few years of effort always looks more daunting looking ahead than behind you. Don't be afraid to make long-term commitments to improve your profession or health care. You can always enlist the help of your colleagues.

I hope that you find these words helpful. You can look forward to a career that will allow you to use your talents in meaningful ways.

Sincerely,

Joseph T. DiPiro
Executive Dean
South Carolina College of Pharmacy

Ken Fagerman

Ken Fagerman, *RPh, MM is currently a practicing clinical pharmacist and manager in the specialty area of infusion pharmacy and adjunct professor of pharmacology. His career experience includes pharmacy clinical, retail and managerial positions. He is a past president of the St. Joseph County Pharmacy Association and leader of a highly successful retail pharmacy crime watch. In 2006, Ken was the recipient of the George A. Cooper Life saving award and was also inducted into the St. Joseph County American Red Cross Hall of Heroes for risking his own life to save the life of an office worker overcome by gas and unable to escape from an office building. Ken went into this dangerous area, restored the victim's breathing (who coincidentally was also a patient of his pharmacy) and maintained her airway while waiting for rescue and she made a full recovery.*

Dear Ken,

It seems to me that today virtually every occupation labels itself as "professional." Why, there are even professional sanitation engineers picking up trash. I think pharmacists sometimes forget that they are professionals once they are out in the working world for a while. I know I cringe sometimes at the way some of my colleagues behave, dress, and act. I am proud of what I had accomplished and my "professional" status now that I am graduating from pharmacy school.

We mistakenly let others beat this pride out of us. Physicians, nurses etc. have a right to also be proud of their training as well but we shouldn't let them beat us down and we, more than ever, need to maintain a professional image. What are the traits of a professional? Here is my opinion and a definition I ran across, added to and that warrants memorizing and repeating.

Traits of a True Professional

1. A professional has specialized expertise and knowledge gained through years of training and experience.
2. A professional maintains that expertise through continued education and training.

3. A professional adheres to professional standards of practice.
4. A professional is ethical and impartial.
5. A professional is professional in appearance (neatly and well dressed, well spoken).
6. A professional exercises good judgment in making decisions in the best interest of the patient and organization.

Think about it. These traits are what separate you from a trade. Be proud of your profession.

Secondly, have you ever noticed how doctors treat each other? Pharmacists need a lesson here. We are our own worst enemies. We speak badly of each other to customers and staff. We use "cut throat" competitive pricing tactics and then we bemoan about our wages, prices and the "state of pharmacy". Physicians also, as a general rule, respect each other's expertise and aren't afraid to confer and refer patients. Pharmacists as a general rule don't do this. Oh sure, when we don't have something we send the patient to another store or might ask a colleague a question but when it comes to using or supporting another pharmacist in a business sense we won't do it.

You will learn this lesson, as you will start a home infusion consulting business. You will find out that your biggest business obstacle to closing the deal was the current pharmacist. These operations fail for a lot of reasons and an inexperienced pharmacist is often a key reason. No, it's usually not for clinical reasons. It's the lack of business skills. Respect and reinforce the success and expertise of fellow pharmacists especially colleagues that have bucked the trends. It is also difficult to run your own business or store, continue to compound or offer a consulting business. Support your colleagues when you can.

Finally, what is the "pharmacist personality" and where do you fit in? You probably already have a preconceived idea. Maybe your family, friends and classmates have already "slotted you in".

Intellectual, geek, all black and white, can't deal with people, you name it, and you've probably heard it already. So who cares, all professions and careers have stereotypes, right?

Well it matters a lot!! That's because people, like it or not, make decisions about you based on those stereotypes and how you reinforce them (see Traits of a True Professional). In fact, they usually exaggerate the things you do or don't do. Yes, just like a pharmacist, get them out from behind the counter and they fall apart.

So what do you do? You compensate. You become aware of your actual or perceived shortcomings. Think you don't have any? Well, try role-playing and videotape how you deal with a problem. You probably won't like what you see. We all "sell" ourselves everyday to our customers, co-workers, friends and even our spouses. Most of the time it's not the message we give but how we deliver it that matters the most and we must continually work on our "delivery".

Particularly effective method that you should use as you embark on your career and especially with a new job is to ask for periodic meetings with your boss. At these meetings, ask how you are doing and if there is anything you can "alter or improve upon". This not only helps you learn your shortcomings but it effectively "clears the air" in terms of any employment issues and for upcoming reviews. Try to develop a mentor relationship with your boss or perhaps another successful pharmacist you admire. Maybe someday you will be the admired, professional, caring and knowledgeable pharmacist we all aspire to.

Good luck as you embark upon your profession,

Ken

Nathan Gabhart

Nathan Gabhart, *RPh is the Vice President of Pharmacy operations for an independent pharmacy chain that employs over 200 people. He began his college education in 1991 at Purdue University. While in pharmacy school, he became very involved in APhA-ASP thanks to Jane Krause, who was a professor. He held several local and regional offices, and went on to become the first national officer in the history of Purdue in 1997.*

He is one of five children and grew up in a single parent family. He began working in a pharmacy/restaurant at 16 years old. It was an old time pharmacy with cards, candy, a restaurant, etc. The pharmacists who worked there would come over and have coffee during the day so this gave him the opportunity to talk with them. This is what began his pharmacy career.

Currently, he serves as District Representative for the Indiana Pharmacists Alliance, as well as serving on the United Drug State Council.

Nathan,

In my opinion everyone has one life to live and it is totally dependent upon that individual on how they live it. They, and only they, know what they are capable of achieving in life. Be cautious on listening to those around you who give you encouragement or discouragement because they can both be equally counter-productive. You have to have that internal drive, whether you call it passion or whatever; you must possess it in order to be successful in life. It is not possible to fake this passion either. If you fake it you will fail. Also, always remember, the amount of money you make is not reflective of your level of success. If you use monetary value as a measuring tool, you will never be completely successful or happy. You must also know your priorities in life. If you only have one or two priorities then it is very difficult to live a balanced life. Whether your priorities are friends, children, your spouse, your job, your religion or hobbies, you must have several and you must also set aside enough time to devote to each of them.

Realize that not everyone knows what they want to get out of life. If you do know, then realize how fortunate you truly are and do everything within your power to achieve it.

Regards,

Nathan

Jane Gervasio

Jane Gervasio *is the Director of Clinical Research and Scholarship for Butler University and teaches as an Assistant Professor of Pharmacy Practice both in the classroom and the clinical setting as a nutrition support pharmacist at Methodist Hospital, Clarian Health Partners. Dr. Gervasio received her Bachelor of Science Degree in Pharmacy from Butler University in 1988, and returned to Butler to complete her Doctor of Pharmacy Degree in 1995. She completed a specialty residency in critical care and nutrition support and a fellowship in metabolic support at the University of Tennessee in Memphis. Dr. Gervasio is nationally recognized for her expertise in the area of nutrition support. She is a board certified nutrition support pharmacist. She speaks both locally and nationally and has published numerous peer-reviewed manuscripts and book chapters on various nutritional and gastrointestinal related topics. Dr. Gervasio serves on various committees for several nationally recognized organizations, including the American Society for Parenteral and Enteral Nutrition, the Society of Critical Care Medicine, and the Specialty Council on Nutrition Support Pharmacy Practice for the Board of Pharmaceutical Specialties.*

Dear Jane,

What an accomplishment! Getting through pharmacy school was challenging. All those days and nights studying, worry if you passed, praying that you make it through—it was all worth it, trust me! And the best is yet to come!

Embrace life's opportunities and believe in yourself. Both in your professional and personal life, opportunities will be presented. Don't let fear or intimidation stop you. Everyone has trepidation the first time they do anything. Remember the first day in college. How scared you were but you made it. You even enjoyed it. Approach opportunities in the same manner. Challenge yourself. Believe in yourself. Go beyond the norm. Trust me, in doing so life will only be more exciting and fulfilling. And know, not everything you do will you do well. That's ok. Regret is only found in not attempting the challenge. Much may be learned from failure. That too is a life lesson. The trick is to keep going, to pick yourself up and keep trying. Have confidence in yourself. You can do it. Or, to quote your nieces and nephews, who in turn quote *The Little Engine That Could*, "I think I can, I think I can."

Be an agent of change. Change will happen. It is inevitable. The world will not stand still. Neither should you. Continue to learn and grow with changes. And you will have change. Your life's path will take you many directions and places. How exciting it will be and what learning opportunities!

Keep balance with all aspects of your life. Your career is important but it is not everything. Your spirituality, your family, your friends, they are your strength. Allow them to support you when you are weak and remember to support them. Enjoy your family and friends. Spend time with them but also spend time with yourself. Remember to take care of yourself. Focusing on your health and your happiness will give you a sense of peace and joy which you in turn can pass on to others.

And lastly laugh, especially at yourself. You will have successes and failures, ups and downs, but in the end, if you can smile through it all, you ultimately will win at this game we call life!

Enjoy the adventure!

Jane

Judith Jacobi

Judith Jacobi, *PharmD, FCCM, FCCP, BCPS works as a Critical Care Pharmacy Specialist at Methodist Hospital in Indianapolis, IN. She graduated from Purdue University with a BS in Pharmacy and from the University of Minnesota with her Doctor of Pharmacy degree. She was one of the first 2 residents at the Ohio State University Critical Care Residency under Joe Dasta and has used that experience throughout her career. She has held positions at Wishard Hospital, and as clinical pharmacy specialist at St. Elizabeth Hospital in Lafayette, IN. She was also a critical care pharmacy specialist at St. Vincent Hospital in Indianapolis before her current position.*

Jacobi was influenced by her grandfather, George Horky, who had a pharmacy in Milwaukee, Wisconsin, and a neighbor, a pharmacist, that introduced her to the idea of women working in this field.

In 1989, the Society of Critical Care Medicine started a pharmacy section and gained a pharmacist seat on the governing Council in 1995. Jacobi assumed that position in 1998, and has recently moved onto the Executive Committee of SCCM – currently as Treasurer, and will become President of that organization in 2010. It is indeed an honor for her to represent pharmacists as a member of the critical care team and to illustrate how that has been validated by this multiprofessional organization.

Dear Self,

Welcome to the world of pharmacy! Whatever you think it means to be a pharmacist is probably only partially correct, and it is changing every day—to be exactly what you want it to be. You are responsible for what you do and who you become through the opportunities you seek and through many that just appear. Take advantage of these experiences. It may not seem important at the time, but in hindsight it may be. As you pursue your training, remember that many people are ready to teach you, if you are ready to learn. Nurses are the most important people in the hospital in many ways. They are closest to the patient and understand their needs intimately. No matter what your job becomes, it is ultimately about that patient, "the guy in the bed". However, every member of the team lends important insight about patient care—seek out their opinions and advice.

As we think about the "guy in the bed" always remember that their needs don't revolve around your schedule. As a health professional, be prepared to support your co-workers and meet the needs of the patient. You can always make a difference.

After learning the hard way, I can strongly advise you to follow your instincts at all times. If you ever have to stop and think "I wonder if I should…?" The answer is always yes if it involves a patient care issue. You never want to regret having missed an opportunity. You will get some flack at times, but never take it personally—it is not about you.

However, there are many things you will need to be a better pharmacist and a better person. It will be hard work to keep up with the burgeoning medical literature (pick a narrow specialty focus if you are smart)—you will never be caught up on your reading! However, always be willing to learn—pharmacists are like bulldogs in our beliefs, and it is difficult to change our minds, but we don't know as much as we should about patient care, and some therapies we thought were great did not stay that way when studied further.

Look for opportunities to join organizations and network with other pharmacists—there is usually a different way to approach a problem and someone else may just have the insight you need. A perspective on the world outside your job is very humbling and sometimes provides essential affirmation. Professional organizations at the local level are a great resource close to home, but the opportunities are immense. When someone wants to participate and speaks up when asked, or more importantly volunteers an opinion before being asked, the doors will open.

There are many important contributions that pharmacists can make in patient care, but our knowledge of pharmacokinetics and willingness to apply these principles to patient care is a truly unique skill. In addition, always focus on aspects of medication safety. That is another very important contribution—analyze everything about what can go wrong with a drug—adverse effects, improper dosing and administration, and suboptimal patient utilization, and plan ahead to minimize these risks.

Finally, remember no one is perfect. If you make a mistake—and you will—be ready to acknowledge it to your boss, co-workers, and the patient or family of those impacted. Learn from the mistakes of others, and many will be better off.

Sincerely,

Judith Jacobi, PharmD, FCCM, FCCP, BCPS
Critical Care Pharmacy Specialist
Methodist Hospital
Indianapolis, IN

Julie Koehler

Dr. Julie Koehler is Chair of the Department of Pharmacy Practice and Associate Professor at Butler University in the College of Pharmacy and Health Sciences. Dr. Koehler earned her Doctor of Pharmacy degree from Purdue University in 1997. In 1998, she completed an ASHP-accredited Pharmacy Practice Residency at the Indiana University Medical Center. Upon completion of her residency training at IUMC, she accepted a co-funded position in Indianapolis with Clarian Health Partners as a Clinical Pharmacist in Family Medicine and with Butler University as an Assistant Professor of Pharmacy Practice in the College of Pharmacy and Health Sciences. Since 1998, Dr. Koehler has served at Clarian as both an inpatient clinical pharmacist at Methodist Hospital and an ambulatory care clinical pharmacist at the Indiana University-Methodist Family Practice Center for the Indiana University School of Medicine Family Medicine Residency Program. For the Department of Family Medicine, Dr. Koehler serves a faculty instructor for both medical residents and medical students. Since joining Clarian in 1998, Dr. Koehler has helped to establish and oversee numerous pharmacist-run clinical services for the Indiana University-Methodist Family Practice Center, including an Anticoagulation Clinic, a Hyperlipidemia Clinic, a Pulmonary Education Clinic, and a Smoking Cessation Program. At Methodist Hospital, Dr. Koehler is also a member of the Pulmonary Rehabilitation Team, and she provides monthly instruction to patients with chronic obstructive lung disease about their medications. Dr. Koehler also serves as the Director of the Family Medicine PGY-2 Residency Program at Clarian, which she established in 1999. Dr. Koehler has served as Chair of the Department of Pharmacy Practice at Butler since 2003. Since joining Butler in 1998, Dr. Koehler has received numerous teaching awards, including the Outstanding Pharmacy Practice Professor Award, the Mortar Board Society Teacher of the Year, and the Board of Visitors Award of Excellence. In addition, she was recognized in 2007 by the Indianapolis Business Journal as a recipient of the "Forty Under 40" Award.

Dear Julie,

Congratulations on selecting the profession of pharmacy as your chosen career path! There are so many rewarding opportunities that lie ahead of you.

I know you're still in school, and you're probably looking forward to getting to the point at which you "know it all." I hate to break it to you, but you'll never reach that point! Don't worry – no one ever does. There is constantly new information being produced: new drugs, new indications, new dosage forms, new studies, new findings, new clinical guidelines. Your challenge will become staying abreast of all that's new. So, while you are still in school, embrace that fact that it's not just about *what* you know. It's also about *how* you obtain the information. To that end, the skills you develop as a learner are invaluable, and your willingness to accept the concept of "life-long learning" is essential. Pharmacy is one of the few professions within healthcare that is such a heavily knowledge-driven field. The practice of pharmacy doesn't require much technical skill. This is somewhat unlike the practice of medicine, in which a physician may need to perform a thorough physical exam or surgical procedure. As pharmacists, we are looked to by our colleagues, by our patients, and by other health care professionals as providers of knowledge. We may be asked questions to which we readily know the answers, and yet, we may also be asked questions to which the answers are not readily known. In such cases, it's our job to find the answers. That's where the continuous learning part comes in!

Don't be afraid to admit when you don't know something. I've been in practice for ten years, and I always tell my students and residents that I still learn something new almost every day. It's perfectly okay to say, "I don't know, but I'll find out!" Never knowingly provide incorrect information. Following this simple rule will help you to keep the respect you probably worked hard to gain in the first place. Breaking this simple rule will likely cause others not to trust you. If you make mistakes along the way, first of all, recognize that you are human – it happens. Second, *admit your mistakes*. Third, *learn from them*. Sometimes the culture of your work environment may make admitting your mistakes difficult to do, but always remember that it's the right thing to do.

While you're still in school and throughout your career, continuously work to improve or develop your communication skills. Your effectiveness as a pharmacist may be highly dependent upon them! When you are asked a question, whether you know the answer, or whether you have to find the answer, in the end you have to be able to effectively communicate the answer to someone else. How you communicate that information depends heavily on to whom it must be communicated. For example, how you might communicate information to a patient may be very different than how you might communicate it to a physician, a nurse, or another pharmacist.

When choosing your first job, think broadly and look to the future. What may be the most attractive offer right out of the gate may not exactly get you where you want to be five or ten years from now. I know it's probably hard to believe, but the best job may not always be the one that pays the most. In fact, don't expect money to bring you happiness. It's not always as much about the work as it is the people with whom you work – and that also includes your patients. There's nothing more satisfying at the end of the day than to go home knowing that you helped or perhaps made a difference in someone else's life that day. No matter how large or small the sacrifice you make, giving feels good, especially when you help those who could not otherwise help themselves. Winston Churchill once said, "We make a living by what we get, but we make a life by what we give."

Even if you are happy in your job, don't automatically close the door when a new opportunity arrives. Throughout your career, take the time to learn about other opportunities as they present themselves. Sometimes, a better opportunity may very well come along, and you'll be glad you didn't let it pass you by. Many times, however, exploring other opportunities along the way helps you realize that the best place for you to be is right where you are today! This can be tremendously helpful in satisfying that curiosity inside you, which occasionally makes you wonder if the grass is really greener on the other side.

Speaking of opportunity, sometimes it is not always obvious. In fact, sometimes you need to be the one to *create* an opportunity that doesn't currently exist. This takes courage and perseverance, believe me it's not impossible. It was Ralph Waldo Emerson who once said, "Go not where the path may lead. Go, instead, where there is no path, and leave a trail." Sometimes you need to create a new path for yourself and for others to follow.

Develop an understanding of and an appreciation for the roles and responsibilities of other healthcare professionals who work outside of your discipline. It takes a qualified team of healthcare professionals to provide the best possible care for a given patient, and that team may include a physician, a pharmacist, a nurse, a physical therapist, a social worker, and perhaps others. Chances are, you'll be collaborating with others in the field when caring for patients.

As you continue throughout your career path, no matter what, don't burn bridges along the way. You'll be surprised how many times you might have to cross the same river, so to speak. The world of pharmacy is very small, and you'll be shocked to know how many people within our profession actually know each other or are connected in some way. This never ceases to amaze me.

No matter what area of pharmacy practice you choose, put your heart into it. Abraham Lincoln once said, "Whatever you are, be a good one." Give your best to your employer, to your colleagues, to your patients, and to all those whom you serve. It's one of the best investments you can make!

Best wishes!

Julie

Lindsay Koselke

Lindsay Koselke *is a Clinical/Staff Pharmacist at St. Anthony Memorial Hospital, where she works as a pharmacy liaison for the open heart surgery program and as an Affiliate Associate Professor of Clinical Pharmacy through Purdue University (serving as a preceptor to students since 2005). She is also in the process of instituting an Anticoagulation Clinic at St. Anthony Memorial Hospital in Michigan City, IN. Lindsay graduated from Butler University with her PharmD in 2002.*

Dear Lindsay,

I am writing to you today to give you some guidance and advice on your upcoming pharmacy career. It has been five years since I received my degree, and I still truly enjoy my chosen profession. I would definitely choose to do it all again if given the opportunity.

My first piece of advice is that you be open to all employment opportunities. I worked in the retail setting all through college so that had become my comfort zone. Upon graduation I was intimidated by the hospital setting and did not feel I had the skills or knowledge required. I accepted a position with a retail organization, and though I enjoyed my work, I soon felt the need to pursue other opportunities. A position arose in the hospital setting and I decided to take a chance and test my abilities and confidence in myself. I am happy to say that it was definitely the right decision. The hospital setting has allowed me to become involved in so many different programs and situations. I never should have doubted my abilities. So my message to you is not to limit yourself. Have confidence in your abilities, and do not be afraid to push yourself to find a position that motivates and challenges you every day!

Another piece of wisdom I have for you is to never be afraid of change. Technology is moving so fast that it seems the workplace is different every single day! When I started at the hospital the physician orders came to the pharmacy as a yellow triplicate form, and one of the sole jobs of the pharmacists was to enter the orders into the computer. Now the physician orders are scanned to a location 60 miles away to be entered, thus allowing our pharmacists to be on the medical floors doing more and more clinical work. It has been a huge transition for our department. Many of the pharmacists are hesitant of the process and are scared to enter into this whole new realm of pharmacy. Once again it is a wonderful opportunity to step outside the comfort zone! Working with an open heart surgery program and developing an anticoagulation clinic have become much more professionally fulfilling to me than simply entering orders all day.

My final suggestion for you is to become involved in a preceptor program. Working with pharmacy students every day is a fantastic way to keep your knowledge base and skill levels current. I feel that I learn as much from working with them as they gain from the rotation experience.

Best of luck to you!

Lindsay R. Koselke, PharmD
Staff/Clinical Pharmacist
St. Anthony Memorial Hospital, Michigan City, IN
Butler University Class of 2002

Lucinda Maine

Lucinda Maine *serves as Executive Vice President and CEO of the American Association of Colleges of Pharmacy. Dr. Maine has also served as a senior staff member for the American Pharmacists Association (APhA) with responsibility for policy, communications, and volunteer development. She held academic appointments at the University of Minnesota and Samford University. At Samford she served as Associate Dean for Student and Alumni Affairs. Dr. Maine is a pharmacy graduate of Auburn University (BS 1980) and received her doctorate at the University of Minnesota in 1985. Lucinda has been active in leadership roles holding the positions of Speaker of the APhA House of Delegates and APhA Trustee. In 2004 Lucinda was recognized by the APhA Academy of Students of Pharmacy with the Linwood F. Tice Friend of APhA-ASP Award. She also received the Prescott Leadership Award from Phi Delta Chi and the Bowl of Hygeia.*

Dear Lucinda,

What a fabulous beginning to the career in pharmacy that is about to unfold in ways you cannot begin to predict today. Remember that you used to think you would be a first grade teacher and now look where life has taken you! Keep your mind open to all the wonderful possibilities that exist in the profession of pharmacy, which you've clearly come to love.

You made a great decision already to continue your formal education in a graduate program at University of Minnesota. Although not everyone needs additional education or even a residency, advanced education clearly accelerates learning and offers you another credential to distinguish yourself in the marketplace. It also opens some doors, like academia and some positions in management that might not open as easily without advanced education.

There are so many choices for the career path ahead. I am so grateful that there is a program that can help me identify where my interests and career preferences might best align with pharmacist positions. APhA's Career Pathways Program (http://

www.aphanet.org/pathways/pathways.html) is worth spending some time on to help better appreciate how to make a good decision on selecting my first post-graduate position.

Don't be afraid to experiment a little, too. What I mean by that is during the years you are in graduate school try different positions for short term experiences. Summer internships and things like that are the best way to gain new insights into the work that pharmacists actually do. You will likely have three to five different mini-careers if you are like most workers today. Don't hesitate to try something really different from earlier experiences because it may all add up to a truly fabulous opportunity down the road that draws upon all the different skills you gain in each earlier position.

One thing I strongly recommend you do is answer the question: What is truly my goal for my work in pharmacy? I think for me the general answer was to improve the way medications are used in the health care system. There is so much room for improvement! That is job security in a big way!

Finally, seek balance in your life. Work is important but so too is family and community. You are a well-educated person with a strong commitment to help others!

Best always,

Lucinda

Susan Malecha

Susan Malecha, *PharmD, MBA, a has over fifteen years pharmaceutical industry experience in new product development, managed markets, medical education, medical affairs, medical science liaison and management experience. She graduated from Butler University, completed her Doctor of Pharmacy at University of Illinois at Chicago, and earned her MBA from Keller Graduate School of Management. She is currently a Senior Medical Science Liaison National Manager at Genentech for two field-based Medical Science Liaison teams, Dermatology and Endocrinology. Recently, she was Senior Director, Field Based Liaisons at InterMune, Inc., a biopharmaceutical company focused on developing and commercializing innovative therapies in pulmonology and hepatology. Prior to her biotech positions, Susan was the Director of Managed Care Clinical Managers and the Director of Medical Education at Abbott Laboratories, the Director of Cardiovascular Applied Therapeutics at Searle, and held various positions in Medical Information and Drug Safety at Boots Pharmaceuticals. She has held adjunct faculty positions at University of Illinois at Chicago and Midwestern College of Pharmacy. She is an active presenter/lecturer on Medical Affairs topics for pharmaceutical industry. She participates on the Board of Visitors for Butler University College of Pharmacy and Health Sciences and on the Editorial Board for the MSL Institute. She has published papers in Pharmacotherapy, Drug Information Journal, MSL Quarterly, DIA Forum, and American Journal of Pharmaceutical Education.*

Dear Susan,

You are so fortunate to have the tools for a satisfying and financially rewarding career in pharmacy. It is important for you to keep those tools sharp and in use.

As we learned in pharmacy and medical training, biological atrophy is the partial or complete wasting away of part of the body. Some causes include poor nourishment, poor circulation, loss of nerve supply to the target organ, loss of hormone function or support, or lack of exercise. What is career atrophy?

Simply put, it is a wasting away of some talents and skills that could be used, but are not, due to job circumstances, company policies, lack of opportunity, or other personnel reasons. Depending on the precipatory cause, this could lead to a state of dissatisfaction, malaise, confusion or burnout.

Are pharmacists susceptible to career atrophy? Of course! Every career has the potential for this ill effect.

Many competencies are required to be a successful pharmacist in various settings, including the biotech/pharmaceutical industry. These include therapeutic knowledge, strategic thinking, technology skills, influencing skills, regulatory knowledge, clinical development understanding, business knowledge, creativity and project management. In some work setting situations, depending on therapeutic area supported, stage of product development, guidelines and policies set by the company, management, or other reasons, some of the skills are not utilized consistently. Loss of skill use can lead to career atrophy.

So what can one to do to prevent atrophy?

Exercise and mobilize.

It is important to maintain sharp skills by keeping them in use. In the unfortunate event that the current pharmacy position does not offer opportunities to employ these skills, take advantage of situations that will. Offer to be part of a collaboration project with business colleagues. Ask to attend seminars in areas of interest for professional growth. Put yourself in situations that require fine tuning of competencies and constant learning. Ask management for specific help in finding opportunities.

If the job still does not provide a means for learning and proficiency development, outside organizations, such as Drug Information Association and Society of Clinical Research

Associates, to name a few, provide courses and avenues for career growth.

If nothing else, stay flexible, be aware, and keep learning new technologies. In this industry, what is hot today can be obsolete tomorrow. In whatever situation you find yourself, keep learning. Perhaps another career option will appear.

Career atrophy can happen to anyone. It is up to each person to assess their attainable career condition. This can be done by measuring oneself against a gold standard of accomplishing a balanced life, utilizing abilities to the fullest and seeking fulfillment with one's intellectual course of professional occupation. An optimal career state is similar to balanced physical and mental health. Individuals need to alter a career or job when desirable or necessary to maintain satisfaction or harmony between who they are and what they do.

Remember, exercise and mobilize,

Susan

Rhalene Patajo

Rhalene Patajo *is Manager of Regional Medicine for Boehringer Ingelheim Pharmaceuticals, Inc. Her pharmacy experience includes retail pharmacy, managed care pharmacy, and pharmaceutical sales. She graduated from the University of Washington School of Pharmacy in 2000 with her PharmD degree and enjoys being continuously involved at her alma mater as guest faculty.*

Dear Rhalene,

Congratulations! You should be so proud of yourself right now! You worked so hard in school and now you can get paid. Now the real fun of life can begin…

I know you're also scared. Pharmacy school was fun, wasn't it? You got to learn, have a great time connecting with everyone in your class, and you tried so hard not to get confused with all that memorizing you had to do in biochemistry and medicinal chemistry. Remember how you always told yourself that you were there because you wanted to help patients? That's exactly the BIG PICTURE I want you to remember, whatever you decide to do or be in pharmacy. You are going to do great things!

And don't get so wound up that you don't feel like doing a residency or loving only institutional pharmacy. The beauty of this profession is you can do whatever you want. What is your niche? Find it! So you want to be out there with the patients? Then do retail or do managed care and have fun doing it!

I know you have been thinking about the pharmaceutical industry. You know, you can do that, too! There are so many avenues you can take, so just know that you are making the best decision for you and what you want out of life and what you want for work-life balance. Remember that word "balance" because that will be what you will want in anything and everything you decide to do with your profession and with your life.

So what will you learn now when you are out of pharmacy school? Flexibility! There is so much you can do! There is so much that you can become with your pharmacy degree… whether you want to work in a pharmacy or in the industry or elsewhere. AND…if you want to have a family and want to be a working mom, you can have that, too! I wouldn't trade anything to go back to school and change what I went to school for. I still believe that I am helping patients as a medical liaison, trying to get the best drugs out there with the best data to support it. I love what I do and hope the best for you…find something that you will love within this great field! Good luck!

All the best,

Rhalene

Amanda Place

Amanda Place *is currently pharmacy manager and co-owner of Access to Care Pharmacy, LLC, an outpatient pharmacy focused on advancing patient education and clinical outpatient practice. She also is involved with pharmacy student education, precepting PharmD candidates on rotation and serving as an Adjunct Assistant Professor of Pharmacy Practice at Butler University's College of Pharmacy and Health Sciences. She received her pharmacy degree from Butler University. While she has had exposure to other pharmacy disciplines, outpatient care has always been and continues to be her major focus and passion.*

Dear Amanda of the past,

Thank goodness you are done. Pharmacy school was an uphill road, since you worked full-time like so many students do to make ends meet. Now that you are done you can focus on the goals that inspired you to go to pharmacy school in the first place: patient care and drug knowledge. Stepping forward into your new role as pharmacist will be exciting and demanding but ultimately rewarding. This transition can be a challenge for some young pharmacists; however, here are some thoughts to help keep you going in the right direction.

Don't ever let your patients become an "Rx number" or a "case". Unlike your work in school, these are real people with all the complexities that come with being human. Even when you are more frustrated than you have ever been, remember this. Take the time to be kind, compassionate, and thorough (even when you "know" the answer), and you will be rewarded time and time again-sometimes from the most unexpected sources.

Love your job or move on! Life is too short and the list of opportunities too long to waste it in a job you don't love. I'm not talking about the down days that we all have once in a while. But if you don't have an overall sense of purpose, accomplishment, and vision that gets you going in the morning and keeps you smiling at 5 minutes after closing, then get out.

You might be scared to take the leap, but you won't look back and wonder what you did with your life or what you have to show for your years of service besides the decorative company clock. Your love for your job will leave you more richly satisfied than any long-term but ambivalent work history could.

And finally, take that leap! Whether it is a professional or personal leap, or even one that fits both labels, do it. The lows will be lower, but the highs will be higher and your regrets won't be "I wish I had tried…". You are in the position to impact so many people in a positive way; don't let your inhibitions keep you tied to the conventional and expected careers if your path is in a different direction.

In some cases, the cliché is also true…being true to yourself and devoted to your patients is the best guiding principle you can have. Keep this in your mind and heart and you will be proud of the pharmacist and person you become.

Lovingly,

Amanda of the future

Jeffrey Rein

Jeffrey A. Rein *is chief executive officer for Walgreens, the nation's largest drugstore chain. Rein joined Walgreens as an assistant store manager in 1982, and over a 25-year span, was promoted to a store manager, district manager, divisional vice president and treasurer, executive vice president of marketing, president and chief operating officer, and chairman and CEO. Rein earned his accounting degree in 1974 and his pharmacy degree in 1980, both from the University of Arizona, Tucson. Rein serves on the board of directors for the National Association of Chain Drugstores (NACDS), Midwest Young Artists, Midtown Educational Foundation and the Retail Industry Leaders Association.*

Dear Jeff,

Congratulations! Your schooling in high school and college has helped you develop discipline and has exposed you to work with different people – both great assets in the years to come. But Jeff, your learning has just begun. Here are three things that you'll find valuable in the future.

Your accounting and pharmacy degrees will help you – big time. It might surprise you to know that the college class that you'll find most enriched you was French Literature, which I know you took on a whim. It has helped open your mind to the many, many diverse cultures that you'll encounter over the next 30-plus years. You'll be surprised by some twists and turns, but nothing you learn along the way is wasted as long as you keep your mind open. You must be flexible, tolerant, and open to other opinions. That means you'll have to do more listening. Ask people questions, and keep in mind that your way isn't the only way or even the best way.

Being a nice person will also serve you well in your career. Sure, people will take advantage of you, but being kind and patient with others will help you develop relationships and pull folks together. You'll soon learn that it's the people around you that will bring you success. When you're nice to others, people on the receiving end respect you more.

And finally, take care of yourself. I can tell you that in your 50s you'll be quite a health nut who preaches exercise and eating right, but I wish you would've started the health craze decades earlier. Exercise daily and eat lots of vegetables. This will make a difference later and hopefully keep you healthy enough to see your grandkids.

Thanks and have fun!

Jeff

Denis Ribordy

Denis E. Ribordy *is the founder of Ribordy Drugs, a chain of 26 drugstores in Northern Indiana. He was also president of Ribordy Enterprises, which owned local Hallmark stores. Before he retired, he held many positions, including the director of the Indiana Board of Pharmacy, member of the advisory board of the Butler University College of Pharmacy, and director of the National Association of Chain Drug Stores. He also received numerous awards, including the Lake County Pharmacist Association Pharmacist of the Year Award, Butler University Alumni Achievement Award, and U.S. Navy League Man of the Year. Ribordy is a 1952 graduate of Butler University. Ribordy and his wife have four children and eight grandchildren.*

(The following letter has been adapted from a speech written by Ribordy.)

Dear Denis,

What does it take to become an entrepreneur? It really takes a bunch of four letter words:

- Work: Hard work and lots of it.

- Risk: A willingness to gamble everything to make it happen; to go for broke.

- Guts: A four-letter word for confidence. Really, the guts to try something.

- Help: You must seek advice and listen, but YOU must make the decision.

- Dumb: It helps to succeed if you are not so smart that you already know it can't be done.

- Want: You have to want to do it.

Successful entrepreneurs have enormous self-confidence—the kind needed to tolerate uncertainty. It takes self-confidence to admit you are wrong. This same self-confidence enables many to quit their jobs at America's leading corporations, leaving behind regular paychecks, health insurance, prestige, and profit-sharing plans to venture out on their own. Clearly, much of this confidence comes from ego, but it has to be a certain kind of ego. It is not the "I-am-the-big-cheese" kind of ego. These people are big talkers and poor listeners. Instead, it has to be the "I'm-going-to-do-well" sort of ego. An entrepreneur wants to control his own destiny.

Most people won't try—they are not willing to risk the humiliation for failure. Many more people would succeed if they did try.

A true entrepreneur is a doer, not a dreamer.

Sincerely,

Denis

Byron Scott

Byron Scott *was born and raised in Brooklyn New York graduating from Carnasie High School. He attended historically black North Carolina Central University graduating with a degree in Biology with a Chemistry Minor. He matriculated to Florida A&M University School of Pharmacy earning a pharmacy degree. His professional accomplishments were recognized by Florida A&M University by awarding him Florida A&M University's Top 50 Most Notable Graduates which he so graciously accepted.*

His professional career has spanned the pharmaceutical industry focusing in regulatory affairs, most notably Parke-Davis Division of Warner-Lambert where he was Vice President of World Wide Regulatory Affairs and Pfizer. He has been responsible for FDA approval of several drugs, most notably Cognex®, the first drug approved for mild to moderate symptoms of Alzheimer's Disease and Lipitor®, a cholesterol lowering statin that presently grosses over $13 billion in sales.

Byron,

I have been fortunate to have always been on the management track in the pharmaceutical industry. So few pharmacists seek out pharmaceutical industry opportunities early in their career and instead pursue the highest salaries upon graduating. Many lose sight of the intellectual and monetary growth potential and opportunities to make major contributions to society the industry their degree is based provides. My successful career has inspired me to encourage others to pursue careers in industry through mentoring and speaking to young pharmacy professionals.

Think of your career as journey not a job. Your career is a path with a beginning and end not just a finite point. Know where you would like to go and how you will get there. To get off to a fantastic start, think of your first position as going to graduate school except you are getting paid a salary to learn, not paying tuition to learn. Learn everything about your position and industry that is humanly possible. Some young professionals look upon their first position as if they have completed their end goal at their first stop. I chose to look upon my first position as just the start of my journey. I thought if I could stay up all night studying to do well in class, I certainly could stay up all night to master a project or task that I was paid to accomplish well. One aspect that most young professionals lose sight of is the importance of understanding their business. I have always read professional literature to learn, become competent and knowledgeable enough to speak about my company, their competitors and the industry as a whole. With my years of experience and fortunate success I continue to read professional literature daily.

Also during those first couple of years I always asked questions and took notes. I would ask about acronyms that people would use and write them down on index cards to review them that evening. That would allow me to speak the same language as my colleagues and earn respect quickly.

Take every opportunity to speak to groups, write reports and give intelligent feedback in meetings. My oral and written communication skills have taken me a long way. Great speeches and presentations are recited over and over again, great reports become written documentation.

To truly be successful, I have always recognized the importance of behaviors. We all generally enter the workforce with the same level of technical competence, but why do some individuals move up the corporate ladder and some do not? Generally speaking some individuals master the softer skills or behaviors and some do not. Observe the behaviors of successful individuals within your organizations and model yourself accordingly. Do not sit in your office or cubicle and say that my work will speak for itself.

Take the initiative to walk into unfamiliar situations and introduce yourself. Recognize that there are aspects of successful individuals within your organizations who possess skill sets that you may want to emulate, whether it is the way they dress or the way they conduct themselves. When appropriate and time permits ask them to what do they owe their success. Although you do not want to be like someone else there are always some aspects of their persona that if emulated can be used to enhance yourself. Remember your career is a journey and as such chart your course and apply yourself to the same extent that you did while in college.

Regards,

Byron

Ron Snow

Ron Snow *is the Manager, Professional and College Relations for CVS/pharmacy for the entire state of Indiana, downstate Illinois, and Western Kentucky. He has experience in independent pharmacy, nursing home consulting, and chain drug pharmacy. He has held numerous positions during his pharmacy career, including those of Pharmacy Manager, Pharmacy Supervisor, and Regional Recruiter. He is a 1981 graduate of the Drake University College of Pharmacy, and is a past recipient of the Drake University College of Pharmacy Alumni Achievement Award. He is married to a pharmacist (Linda - a 1980 graduate of the Drake University College of Pharmacy), and has a 21 year old son (Jeff) who is a junior at the University of Southern Indiana.*

Ron,

Congratulations on your graduation from pharmacy school. You are entering a great profession, one in which you will leave a lasting impact on the lives of the patients you serve. Please do not take your role lightly. To assist you in your transition from student to pharmacist, I would like to share with you some advice that has served me well during my career;

NEVER LET AN ANGRY PATIENT UPSET YOU – Trying to keep a positive attitude when you have a confrontation with an angry patient is difficult. We have a tendency to let an angry patient get to us and ruin an entire day. I encourage you to focus on those patients who appreciate what you do, which is the overwhelming majority of those you serve.

ALWAYS BE NICE – Make it a priority to always be nice to patients/other health care providers, whether they return that kindness or not. You can turn many difficult and confrontational meetings with patients/health care providers into a professional and meaningful conversation if you remain calm and collected. Nothing is accomplished by treating anyone rudely.

NEVER UNDERESTIMATE YOUR IMPACT ON PATIENTS' LIVES – It is important to know that you have a huge impact on the lives of those you serve, even though you might not recognize it at the time. A situation arose in my career where I had the privilege of providing pharmaceutical care to an infant who was born with multiple medical issues. Four years after the child was born and all her medical problems were behind her, I received a letter from the mother thanking me for the great care I provided her daughter while she was ill. The letter went on to say that she attributed the fact that her daughter was alive today because of the care I provided. I had no idea of the importance that the mother placed on my actions - actions which I thought were ordinary. Never forget that you will be performing ordinary miracles every day.

GIVE BACK TO YOUR PROFESSION - You are blessed to be part of a profession that is professionally and personally rewarding. You should feel some responsibility to ensure that it stays strong and healthy. Giving back can take many forms (become active in pharmacy associations, precept/mentor pharmacy students, financially support your alma mater, etc.), and you should do it in a manner that makes you feel satisfied.

I hope that you take these small pieces of advice and put them into practice. I believe you will have a much more rewarding and enriching career.

I want to wish you good luck as you enter the profession. Please make sure to have fun and serve your patients well.

Sincerely,

Ron

Bill Sonner

Bill Sonner is a Divisional Director of Pharmacy Operations for the Walgreen Company in its Southern Division. After spending his early years growing up in rural North Central Indiana, Bill obtained his BS in Pharmacy graduating from Butler University in May of 1982. Passionate about retail, he began his career as a relief pharmacist for the Ribordy Drug Company in NW Indiana that June. In October 1985, after becoming a Pharmacy Manager and Asst. Store Manager for Ribordy, Bill continued his career with the Walgreen Co. following its acquisition of the Ribordy chain. After many years of pharmacy management in several Walgreen locations, he was promoted to District Pharmacy Supervisor in NW Indiana. Having completed five successful and fulfilling years as such, he was given the opportunity to become one of Walgreen Company's Divisional Directors in March of 2006.

Dear Graduate,

A few days ago one of your classmates, whom I've also had the fortune of mentoring, heard I was coming to Indianapolis for your annual career fair events and dropped me a note explaining how unbelievable it is that graduation is just around the corner. This certainly is a time of great accomplishment and excitement. What's equally or even more exciting is what lies ahead.

As you move toward and through graduation, you will begin your transition from student to licensed professional and healthcare provider. Whichever path you choose to take and wherever you choose to apply your wonderful education, there will be high expectations, challenges, and endless opportunities. Remember the latter because the first two facilitate the third.

Having graduated 25 years ago, I often reflect on where I've been, what path I want to continue following (both personally and professionally), and what was truly important along the way thus far. Now that you've just about completed your education, I encourage you to reflect as well. Not only is it important to look back, but also to glimpse into your future.

Professionalism: What drew you to this school, this profession? I trust you've prioritized this in such a way that you realize you are now very well equipped and charged with helping improve the quality of others' lives. You are also positioned to readily influence others. Lets face it; you have chosen a profession where you very likely will be the most accessible healthcare provider in your community. There is much more to your role as a professional and as a pharmacist than what you do on a daily basis with your patients. You will quickly realize how influential you can be in the aforementioned community.

Flexibility and Attitude: Never in my wildest dreams could I have imagined that this profession would change so much over the course of my career. When I started back in 1982 there was no such thing as computer down time, technical issues, third party payers, formularies, or prior authorizations. Basically, all we had just 25 short years ago was a typewriter, calculator, and file cabinets. Now look at us. Expect two things: change and more change. My point is, as a professional you will be looked to for leadership during times of change. It's easy to take a "what were they thinking when they rolled this out?" attitude. Your job as a leader is to prepare others for, and to facilitate change. This by no means suggests that you never question the intent and the impact of a new process or procedure. If we all think the same, there's not much reason for more than one person to be on the job. Disagreeing can be constructive. All I ask is that you are always ready to offer up a solution when you don't agree with what's been given to you pertaining to a job.

Mentoring and Leadership: I know you've been asked to always be willing to help someone along the way. Mentoring a student, a technician, or a new pharmacist has always been one of the most gratifying activities I've ever taken part in. But don't stop there. Once you have someone "under your wing," don't let them slip. Mentoring is ongoing. Once your student has mastered some of the elementary tasks the next step is to lead them to greater heights. Remember, leadership and success have to do with getting ordinary people to do extraordinary things!

Outreach: Two words describe outreach. Give back. As I mentioned before, you will be positioned to be accessible. With your degree, positioning in the community, and the respect that goes along with both, you will have the opportunity to reach out to others. It's a busy world and a busy work place. It's very easy to get caught up in both. I encourage you to step back and take some time to get involved. Whether it's related to pharmacy, faith, or some other passion, plan for it and make it part of your routine. You probably don't realize how influential you will be someday. Watch for opportunities and seize the moment.

Balance: By now you're probably thinking, "I just want to get this diploma and get working." I thought the same thing. And work you will…and work, and work, and work. The next thing you know, time and other opportunities have passed. As you can see by my bio, I've always been a person to take the opportunity to advance professionally. Don't forget to advance personally as well. I challenge you to find balance between work, family, friends, and personal time. If you do this I guarantee successes in your professional life will be even more rewarding.

I'm thankful to have the opportunity to talk with you today. Hopefully something I've said strikes a chord with you. Congratulations on your tremendous accomplishments. Enjoy the moment and don't forget…always have fun and enjoy what you do. Each day when you wake up, you'll have the chance to make a conscious decision whether you're going to have a good day or a bad one. I'm confident you will choose to have a good one.

With the utmost sincerity,

Bill

Marilyn Speedie

Marilyn K. Speedie, *PhD is dean of the University of Minnesota College of Pharmacy. In this capacity she has the responsibility of building a strong college with balanced mission in a changing health care environment. Her leadership interests include developing the college's education, research and clinical practice missions and working with the profession to enhance pharmacy's ability to meet the health care needs of the state. Particular interests currently include clinical scientist training, pharmacogenomics, rural health care, pharmacy workforce and interprofessional education and practice. Dr. Speedie received a BS in pharmacy in 1970 and a PhD in medicinal chemistry and pharmacognosy from Purdue University in 1974. Before coming to the University as dean and professor of pharmacy in January 1996, she spent 21 years on the faculty of the University of Maryland School of Pharmacy where she last served as professor and chair of pharmaceutical sciences.*

Dr. Speedie has spoken and written extensively on a variety of issues in pharmaceutical education and science. She has been recognized with several teaching awards during her career. Her teaching interests include herbal medicinal agents, antibiotics, infectious diseases, and biotechnology and she has coauthored a textbook and published numerous chapters and articles in these areas. She also is author or coauthor of over 60 scientific and professional articles in the areas of microbial biotransformation, regulation of antibiotic biosynthesis, production of recombinant proteins, and enzymology. In 1993, she was awarded the Paul Dawson Award by the American Association of Colleges of Pharmacy for her research and teaching in the area of biotechnology. She was made a Fellow of the American Association of Pharmaceutical Scientists in 1996. Dr. Speedie served as president of the American Association of Colleges of Pharmacy from July 2006-July 2007 and is currently serving as immediate past president. She is active in other scientific and professional organizations, including serving a four-year term on the Commission on Credentialing for the American Society of Health-System Pharmacists. She has received the Harold R. Popp award from the Minnesota Pharmacists Association and the Hugh Kabat Award from Minnesota Society of Health-System Pharmacy.

Dear Marilyn,

Congratulations on your graduation from the school of pharmacy. As you become a full-fledged member of the wonderful profession of pharmacy, I would like to share with you my experiences and some of the lessons I have learned.

First of all, this profession presents you with many, many opportunities from which to choose and I know you may be facing some indecision about whether you want to provide patient care or teach future pharmacists or do research or some combination of the above. Decide what you most like to do day to day. Do you most want to work with patients and derive the satisfaction of helping them use their medications better? If so, find a location and a pharmacy where patient care is the primary mission. I was fortunate enough to find one where patients' records were kept – way back in 1970 when few community pharmacies even thought of such a thing – and where the pharmacists were the first line of support for the health of the community's residents. Or will you get the most satisfaction from being a faculty member, doing research and teaching? If so, I can promise you that the extra years involved in education will fly by and you will get great satisfaction from being an expert in whatever field you choose. The combination of teaching and research is wonderful in that there is always the satisfaction and stimulation of working with students and helping them learn, even when the research is not going terrifically well, but, on the other hand, nothing is as exciting as when the research is going well and you discover new knowledge about how things work.

Your choices do not have to be exclusive. You will be able to work with students as a preceptor if you decided to go into practice and you will be able to stay connected to the profession if you go into teaching and research. It may take a little extra work, but the rewards are immense. Try to stay connected to the big picture of health care and pharmacy through professional societies or other means. Pharmacy is a continually changing profession with enormous potential to improve the lives of patients everywhere, but it will take the efforts of all of us to achieve recognition of our contributions and potential contributions by the public and by those who pay for health care. Be a part of leading the changes you want to see. The profession needs leaders and there is great satisfaction to be derived from envisioning a change and leading others to implement that change.

Your choices also do not have to be permanent. I have observed many pharmacists who reinvent themselves every ten years or so. Tired of community practice? Try hospital or managed care. Feel a need to learn more? Go back to school – you will be welcomed. There is no excuse for any pharmacist to be bored. Perhaps there is one role that will fit you best while you are having your children and another that you can pursue once they are in school. Stay flexible and keep learning. The changes, while sometimes stressful, will definitely keep you young. Have fun! If the job you are in doesn't leave you feeling satisfied and feeling a sense of fun along with the responsibility, then it is time for a change. Figure out what it is that gives you that sense of fun.

So…can you do everything? Usually I find that people who have asked me that question are really asking, "Can I have a profession and also have a husband and children?" The answer to the latter question is a resounding "yes", but it is also qualified by "you have to make some choices." You are not a superwoman and as you know, you are a bit of an introvert and will need some "alone-time" to recharge your batteries. Recognize that raising a family and pursuing a full-time career will occupy your time. There are other activities you won't be able to pursue so you must decide that that is OK with you; i.e., that some balls you are juggling can hit the floor without causing damage to your sense of self-worth. A clean house? Hire some help. Being a "perfect" parent? Probably not possible (but probably not possible even with all the time in the world). A community volunteer and hobbies? There will be more time as the kids grow up.

The world is your oyster. You have accomplished a great deal already and that accomplishment places you in the enviable position of having many choices of what to do with your life. So, don't be overwhelmed by the choices, but rather choose a path, knowing that there is no one "right " path and satisfaction is awaiting you along any number of them.

Good luck! Keep learning about yourself every day and share those learnings with a younger generation. Someday someone will be asking for your opinion about their choices, someone will be asking you to be a mentor, and you will respond as if your whole life was well thought-out and planned, when, in fact, it is much clearer in retrospect. And you will say, "What luck I had in finding my way here!"

Enjoy the journey,

Marilyn

Leticia Van de Putte

Senator Leticia Van de Putte, *a pharmacist for more than 28 years, represents a large portion of San Antonio and Bexar County. A former five-term state representative, she is now serving her fourth term as a Texas State Senator for District 26. In 2002, she became the Chair of the Senate Hispanic Caucus and is currently the Chair of the Texas Senate Democratic Caucus. Nationally, Senator Van de Putte will serve as co-chair of the 2008 Democratic National Convention. Senator Van de Putte is a member of the National Hispanic Caucus of State Legislators, where she served as President from 2003 to 2005. She has also been actively involved in the National Conference of State Legislatures. She served as a member of the Executive Committee for three years and then as President. She was the first Texan and first Hispanic to ever chair the organization. She is now serving as Chair of NCSL's Foundation.*

Dear Leticia,

Congratulations on your graduation from pharmacy school! You should celebrate with family and friends. While it has been a tough journey, you have earned it. Your graduation marks an end to one chapter and the beginning of your new career. In preparation, I would like to share with you a memory from my experience.

Growing up on the West side of San Antonio, I always wanted to be a pharmacist like my grandfather. He was an old time "Boticario" who owned the La Botica Guadalupana, located in San Antonio's historic Mexican business area, which later became "Market Square." He was the individual in my life who embodied everything that I wanted to be one day. It wasn't just the knowledge that he possessed, it was the way I saw people approach him and ask for his counsel. He was doing clinical pharmacy and disease management long before I knew what those words meant. People looked at him with a feeling of such respect, and I recognized that. I also recognized that he, in turn, gave everyone who walked into the botica the respect that they deserved.

Now that you have graduated and attained the proper tools and skills, you ***must*** have the ability to use those tools for the greater good, as my grandfather always did. And while your education has given you those tools, special people in your life have given you a wonderful commodity called character, just like my grandfather gave to me. It is character that will define your personal course. Character and caring will enable you to treat every individual with dignity and respect, and also to use your skill set to help improve your patients' quality of life.

There was an instance early in my career when I felt the most needed and resourceful as a pharmacist, yet it had nothing to do with my ability to discern drug interactions or disease management. One day, there was a young father who walked into my pharmacy. I saw him several times before, and I knew he spoke very little English. He stood off to the side as I finished counseling one of my patients, then he approached me and asked if I had a few minutes. He handed me an envelope and said that he needed someone to tell him what the letter said so he could explain it to his wife. I said that I was actually very busy filling prescriptions for people waiting, but I would be happy to talk to him in about thirty minutes. Then I looked down and saw that the envelope I was holding read "Coroner's Office," and something told me that I should open it now. The document was a letter explaining to this man why his 18-month-old son had died. I realized at that moment that being a healthcare professional meant more than just accessing tools and knowledge. It was about caring; it was one of the most rewarding and yet most difficult things that I had to do as a healthcare professional. To understand that for this man, the only person he felt could interpret the documents and explain them to him in a manner he could understand was the pharmacist in his neighborhood. I cannot begin to tell you what an awesome responsibility I felt as my heart broke for his family.

Yes, you have been given many gifts, including the capacity for understanding and the tools and skills you have attained. But, these mean nothing without character and caring. Please always hold on to the belief, just as grandpa did, that character and caring come first.

Best of Luck,

Senator Leticia Van de Putte, RPh

John Watt

John Watt *received a BS from Butler College of Pharmacy in 1960. In 1961 he was the opening manager for Hooks' [now CVS] 1st 24 hour drug store at 18th and Illinois St in Indianapolis. In 1964 he purchased Indiana Central Pharmacy at the edge of the Indiana Central College [now Univ. of Indianapolis]. Indiana Central Pharmacy was one of the first [if not the first] pharmacy in Indiana to institute professional fee pricing [as opposed to '% mark-up'] for prescriptions [1965], patient medication profiles [1967], patient medication counseling [1970] and prospective DUR/screening for adverse drug/disease interactions [1974]. Indiana Central Pharmacy closed in 1987. In 1988 John received a MS in Hospital Pharmacy [clinical] from Butler College of Pharmacy. Also in 1988 he took a position at Wishard Memorial Hospital as a clinical staff pharmacist. In 1990 he became clinical staff pharmacist for specialty surgery wards and pioneered pre discharge patient counseling and what is now known as medication reconciliation at discharge for medicine patients at Wishard [after OBRA '90 took effect this became an outpatient pharmacy function]. In 1996 he organized and managed the home IV therapy program for Wishard's indigent patients after the specialty surgery wards were integrated into the general med/surg wards. He became a member of Wishard's Acute Care for the Elderly [ACE] team when it was formed in January 1998. In 2002 he became the consultant pharmacist for the SNU unit of Wishard's ECF. He continues in the last three positions while working part time in preparation for retirement.*

Dear me,

You've now been in school for 18 years or more and have completed one of the more difficult curriculums. You are now entering the 'real world' and need to adjust your outlook for the new realities you will face.

You have learned all about evidence based medicine but you need to keep in mind that this is just a guideline for the most effective treatment of a disease in an ideal patient. Not all patients are ideal. Patients are all different and many of them do not fit neatly into the pigeon holes that evidence based medicine tries to fit them in. You must temper evidence based medicine with what I like to call patient based medicine.

Don't think you cannot practice clinical pharmacy in a community setting. It was hard years ago but it is getting easier in central Indiana now that a computer network is being developed that will allow ever more health care professionals across the region to access information from a wide variety of providers in the area. In a neighborhood pharmacy you will have to use your knowledge of diseases and disease progression to make clinical estimates about your patients whose information is not fully available on the network BUT it can be done.

Should you choose community pharmacy, I hope you enter a practice that allows developing a relationship with your patients. This type of relationship can be very fulfilling if you allow it to be. You are the health care provider that sees them most and often the first to see them in a time of need. Your patients will look up to you as a friend and a learned adviser. You will have an obligation to treat them fairly and with appropriate care.

Above all you must be true to yourself and maintain your personal and professional integrity. "Once you lose your integrity the rest is easy".

God Bless

Hanley Wheeler

Hanley Wheeler *is the Senior Vice President of Central Operations for CVS/Caremark. He currently has responsibility for 2,800 pharmacies in 14 states. He graduated with a BS in Pharmacy from Ohio Northern University in 1982 and began a career as a pharmacist with Revco Drug. He has held numerous management positions with Revco and now CVS, having lived in Cleveland, Dallas, Albuquerque, Charlotte, and now Indianapolis. He is married to his college sweetheart (Mary Murphy Wheeler-ONU class of 82 as well) and is the proud parent of one daughter (Bridget Anne Murphy Wheeler) who is a junior at Indiana University.*

Dear Hanley,

I would like to begin by congratulating you on selecting such a wonderful profession and graduating from pharmacy school. You were very fortunate, recognizing that your choice of pharmacy was based on the fact that one of your high school friend's older brothers had chosen that profession. Fate!

Remember, just because you now have a degree in pharmacy, this does not mean that you know everything. There are many great pharmacists that you will come into contact with. You can learn the practice of pharmacy and you can learn how to become a leader. I encourage you to look for opportunities to be a leader. Make the people that you work with better. Make a positive impact within whatever team you become part of. Thank them.

I know that you have aspirations to climb the "corporate ladder". You should always remember your roots as you do so. Having said that, do not be afraid of a challenge! Move where the opportunities present themselves. Treat every move like a new challenge and adventure. America is a great country! America is a very large country. Try to experience as much of it as you can. It will only make you a more rounded leader.

Make sure that you give quality time to your family. It is very easy to get tied up in your professional goals and miss this opportunity. Enjoy life! Have fun!

Finally, be true to **yourself** and be true to your values. If you do so, you will gain the confidence and respect of the people that are around you at work and at home.

Good luck!

Letters from Contributing Students

Alisha Broberg

Alisha Broberg *is currently a PharmD Candidate at Butler University with expected graduation in May 2008. She has worked as an intern at a retail pharmacy for the past three years. Alisha is involved in numerous organizations, including Rho Chi, Phi Delta Chi, Phi Kappa Phi, and Alpha Lambda Delta.*

Dear Alisha,

Congratulations on making it this far! Graduating from pharmacy school will be quite an accomplishment. The past six years of college have been incredibly challenging and rewarding, but they cannot compare with what is to come. Although you may not be sure what the future holds, it's important to keep in mind that you can make whatever you want out of your life. It's all in your hands.

Remember to always do what makes you the happiest. People will come and go in your life, but in the end, it's only you. You are the only one who knows what is best for you, so don't let others make you feel like you don't deserve it. If you aren't happy, then nothing else matters. Figure out what it is that makes you happy first, and then take the time to share your happiness with others.

Also, remember that the past is in the past. It's time to let go. Consider graduation as your fresh start. Learn from your mistakes, but don't dwell on your regrets. Do not be afraid to make new mistakes; avoiding opportunities will only create new regrets. Step up and put yourself out there. Do something that scares you. In the end you may be surprised with the outcome.

Lastly, don't take anything for granted. This includes your family, your friends, and your life. You are lucky to have such supportive family and friends. Don't ever stop appreciating them. Let them know often how much they mean to you. And live life to its fullest. Right now, the world may seem a little overwhelming—but don't give up on your dreams. Buy your first house. Get married. Be a mother. Travel the world. Love your job. Live happily.

The future holds so much uncertainty, but instead of looking ahead with apprehension, imagine all the possibilities that it holds. Take the advice you have learned from the writers in this book and live the best life you can. You only live once, so there is no sense in living a life that is anything less than wonderful.

Remember, "Everything is always okay in the end. If it's not okay, then it's not the end."

Good luck!

Alisha

Jennell Colwell

Jennell Colwell *anticipates graduating from Butler University in May of 2008 with her PharmD. She has a wonderful husband and daughter, who have supported her through this venture. She is a member of the Indiana Pharmacists Alliance.*

Dear Jennell,

This has been an extremely long and exhausting adventure. Never forget how hard you have worked to get where you are. Remember all the obstacles you have overcome to start your career, and never take your chosen profession for granted. You persevered through college for a reason. Take all of your knowledge and use it to change people's lives.

Also, remember to always thank those around you for everything they do. You can not be successful without great support. Nurses, technicians, social workers, and all those you work with make you a better pharmacist. Especially remember your family. Without their support you would not have been able to spend late nights or entire weekends cramming for clusters or finishing a project.

Balance your life between helping others and being there for your family. One of the many reasons you chose this profession was to help provide for your family's future. You are a role model for S and any other children to follow. You can show them the benefits of going to college and making a career. Along with this, you can also be a great mother to them. You will have to find a way to balance being fulfilled in your career and being the best mother possible. I know you can reach every goal you set, and achieve even more than I can imagine right now.

Congratulations
Jennell

Brad Koselke

Brad Koselke *is a PharmD Candidate at Butler University set to graduate in May 2008. He is currently a member of Phi Delta Chi, Alpha Lambda Delta, Phi Kappa Phi, and Rho Chi.*

Brad,

The end of school is in sight, but as the old cliché says, there is definitely light at the end of the tunnel. Though the last six years of school have been challenging, they have more than prepared you for an incredibly rewarding profession with countless possibilities. With graduation just around the corner, keep the following items in mind as you set out on your career path.

First and foremost, be happy with what you are doing on a daily basis. Many people have taught you the importance of hard work, but at the same time they have also conveyed the importance of living life to the fullest. You're entering a challenging and often stressful profession, so try your best to not become overwhelmed by your daily tasks. Working hard to best serve your patients is key, but working hard to keep yourself and those around you happy will also prove to be incredibly rewarding.

Second, remember that you will soon become one of the most easily accessible health care professionals. Always view your position as an opportunity to give back to the community. You have chosen this career in order to help others, so spend time with your patients and go out of your way to address their needs. Also, personal reward does not have to come in the form of a paycheck. Volunteer. Your time, efforts, and contributions towards helping others will likely prove to be most fulfilling of all.

Finally, don't forget that you wouldn't be where you are today without the help and support from your family and friends. You can never thank them enough for that.

It will be interesting to see where the future takes you, but remember that you have the ultimate say in that matter. Be content, but never stop pushing yourself. The pharmacy profession has so many opportunities available to you, so don't be hesitant to pursue new frontiers. Most important of all, try to make the very best of each and every day.

Good luck,

Brad

Annah Steckel

Annah Steckel*, PharmD Candidate at Butler University, will be graduating from Butler University College of Pharmacy and Health Sciences in May 2008. During college Annah has been involved in multiple organizations including: Kappa Alpha Theta, Lambda Kappa Sigma, IPA, and ASHP. Annah has worked in community pharmacy throughout college and hopes to pursue a residency specializing in ambulatory care upon graduation.*

Dear Annah,

Congratulations on all of your success in the pharmacy field. Doesn't it seem like only yesterday when you were struggling to finish and feeling the pressure of the rest of your life pressing on your shoulders? When you look back and think of that time in your life I hope you look at it with fondness and gratefulness, a fondness for your youth and enthusiasm, and gratefulness for your knowledge and opportunities.

Looking back on your years at Butler University remember the hard work you put into your studies. The long hours in the library spent with books open in front of you trying to cram information into your already overly saturated brain. Remember how your persistence paid off. Remember that your refusal to quit served you well, and that being stubborn about some things in life is a virtue and not a hindrance.

Always remember your mentors, the professors that lead your classroom learning and your fellow pharmacists that lead your professional growth. Remember to always listen, even if you know the answer, because just like when you were a student your mentors and peers still have knowledge to impart and new ways of seeing things that may help to open your eyes to avenues you might not have thought to explore.

Finally remember to always trust yourself; you have a lifetime of experience and knowledge to back you up. But most importantly continue to care about your fellow human beings. Continue to give back to your community with compassion and enthusiasm, and continue to put God's plan before your own as it has served you well in the past and will continue to serve you well into the future.

With best wishes,

Annah

Acknowledgments

Our thanks to:

Dr. Erin Albert for mentoring us through this project.

Mr. Bruce Hancock and Dr. Jane Gervasio for supporting this book idea as our P4 project.

Dean Andritz, Courtney Tuell, Susan Perry, and BUCOPHS faculty for their help in reaching out to pharmacy leaders throughout the country.

Mr. Homer Twigg of Ice Miller and Dr. Robert Holm of Butler University for their guidance and support.

Chris Russell for his continued assistance in making the complex simpler.

Dawn Pearson for her photography.

Professor Gautam Rao of Butler University and his students - Marty Amberger, Austin Athman, Adrienne Bailey, Anna Butterbaugh, Alison Chemers, Kelly Dorman, Rachel Kress, and Mallory Sheets for their amazing artwork and cover design.

Author House for publishing our project.

Johnson & Johnson for providing an unrestricted educational grant towards this project, which allowed each of our 160 graduating pharmacy classmates at Butler to receive a copy of our book in the spring of 2008. Specifically, thanks to Dr. Harlan Weisman and Kathy Curran for support of this project.

Brodie Bertrand and Tiffany Champion for assisting Brad Koselke.

Mike Meginnis for proofreading.

Michael Barbella at <u>Drug Topics</u>, Tabitha Cross at IPA, <u>The Indianapolis Star</u>, and <u>Inside Indiana Business</u> for publicizing our project.

All of our family and friends who have supported us.

Each contributing author for their support in our endeavor: Dr. Erin Albert, Dr. Mary Andritz, Anonymous, Mr. Scott Baker, Dr. Bonnie Brown, Mr. Jack Devine, Dr. Denise Dickson, Dr. Joesph DiPiro, Mr. Ken Fagerman, Mr. Nathan Gabhart, Dr. Jane Gervasio, Dr. Judith Jacobi, Dr. Julie Koehler, Dr. Lindsay Koselke, Dr. Lucinda Maine, Dr. Susan Malecha, Dr. Rhalene Patajo, Dr. Amanda Place, Mr. Jeffrey Rein, Mr. Denis Ribordy, Mr. Byron Scott, Mr. Ron Snow, Mr. Bill Sonner, Dr. Marilyn Speedie, Senator Leticia Van de Putte, Mr. John Watt, and Mr. Hanley Wheeler.

And to all of those serving in the profession of pharmacy for making our profession the most trusted.

Printed in the United States
151073LV00002B/56/P